I0059828

21 WAYS TO REPURPOSE YOUR SOCIAL MEDIA CONTENT

AND WHY REPURPOSING YOUR CONTENT IS CRITICAL

21 Ways To Repurpose Your Social Media Content

And Why Repurposing Your Content Is Critical

Carol Stephen

Copyright © 2020 by Carol Stephen

All rights reserved. No part of this publication may be reproduced, stored in a retrieval system, or transmitted in any form or by any means—for example, electronic, photocopy, recording—without prior permission of the publisher. The only exception is brief quotations in printed reviews.

Printed in the United States of America

About the Author

Carol Stephen has wanted to be a writer since she was a child. At the age of six, she saw some older kids writing in cursive, and she could not wait to learn this mysterious craft. She also read everything she could get her hands on, from the backs of cereal boxes to the library books she pulled home in her little red wagon.

After Carol graduated from U.C. Berkeley with her English degree, she worked as a technical writer for many years. She was a Professional Organizer for close to 10 years and at one point even served as Secretary of the National Association of Professional Organizers (NAPO). She opened her business, *Your Social Media Works*, in 2011. Carol lives near Santa Cruz, California, where she enjoys hiking, the outdoors, and photography.

Find Carol at *Your Social Media Works.*

Dedication

This book is dedicated to my awesome mom, Yoko Nakahira. I cannot say enough positive things about my mom. She was a huge inspiration to me for all my life. She always did the hardest things and made everything look easy. You know how Ginger Rogers did everything Fred Astaire did, only backwards? My mom was Ginger Rogers, although she didn't have the benefit of speaking fluent English. So. To continue the Ginger Rogers metaphor, my mom didn't even have the shoes!

She went through the fire bombings in Tokyo during WWII and escaped to the countryside as a child. She lived through poverty and hardship, but rarely talked about those times. She also lived through racism and misogyny after she moved to the United States. Again, she didn't talk about those events at all. She waved them away like she was shooing away a fly!

My mom had many businesses throughout her life. She owned five Japanese restaurants, two men's clothing stores, a gift shop, and a limo business; and she was an investor in a new bank. All the businesses were located in San Francisco in an extremely competitive environment. Whenever she had a medical procedure done, she never talked about the pain, at all, ever. Hard work never scared her. My mom was a living example of doing the next right thing. She never complained that her life was hard, and she always managed to have fun no matter what.

Mom passed away this year. She will continue to be an inspiration to many. Rest in Peace, mom.

Acknowledgements

This book would not have been possible without the encouragement and editing skills of my best friend, Lisa Eldridge. And I'm very grateful for having Eric Lofholm as my business coach and for his coaching class, *How to Write a Book in a Day* (although this book took far longer than one day).

CONTENTS

INTRODUCTION

This book offers you 21 different ways to recycle or reuse your content on social media. If you want to, you can probably read through this entire book in two or three hours, and afterwards use it as an occasional reference. Or maybe you'll dip into it here and there as the topics interest you. If you feel better able to recycle your articles and other content on social media after reading this book, then I will have met my goal!

WHY RECYCLE YOUR CONTENT?

There are many reasons to recycle your content. First of all, recycling is good for the Earth. Second, according to Hootsuite, "By creating content in different formats, you'll reach more audiences. The more variations you create, the more audiences you'll reach, and the greater chance your content will be shared." Repurposing a blog post into a Facebook video, for example, "could lead to a 62 percent increase in engagement (compared to sharing a post with a photo)." Third, you're a busy professional! Recycling your content saves you time, energy, and helps boost your social media presence—all from the comfort of your own home or office.

Here are a few more reasons to recycle your content:

- Not everyone will see all your awesome writing the first, the second, or even the third time you post or publish it.

- You can approach the topic from a slightly (or radically) different viewpoint when you repurpose your content.

- People quickly forget what they've read or seen, especially online. If they don't remember having seen it before, it will be new to them!

- You may have newer friends or followers who didn't see your content the first time you posted it.

If you would like to connect with me beyond this book, I offer both consulting services and a done-for-you business.

If you would like to read my previous book, it is available on Amazon: *21 Ways to Total Social Media Engagement: That Will Make You Look Like a Pro.*

I have developed many of these ideas over 10 years of engagement on behalf of my clients on social media.

A lot of free content is available on my blog at yoursocialmediaworks.com.

I would love to connect and engage with you on social media. Let's connect at:

http://www.facebook.com/yoursocialmediaworks

twitter.com/Carol_Stephen

www.linkedin.com/in/carolstephen

https://www.pinterest.com/YourSoMeWorks/

https://www.instagram.com/carolstephen/

Your Website Is the Torso of Your Marketing Efforts
(Get started by creating an article or a blog post on your website.)

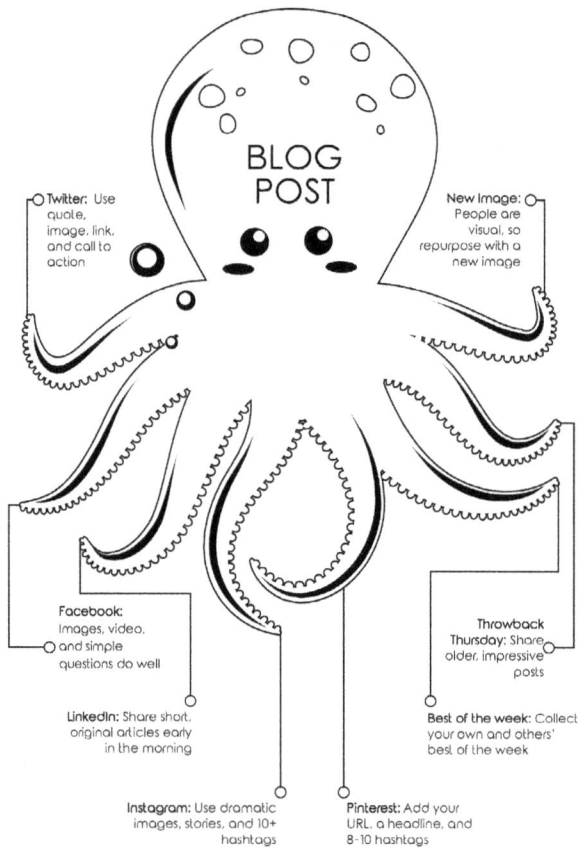

BLOG
POST

Twitter: Use quote, image, link, and call to action

New Image: People are visual, so repurpose with a new image

Facebook: Images, video, and simple questions do well

Throwback Thursday: Share older, impressive posts

LinkedIn: Share short, original articles early in the morning

Best of the week: Collect your own and others' best of the week

Instagram: Use dramatic images, stories, and 10+ hashtags

Pinterest: Add your URL, a headline, and 8-10 hashtags

So what's the best way to get started repurposing your social media content? In my experience, the best content to start with is a blog post or another article from your website, because you will always own that content. (Note: In this book, *blog posts and articles* are used interchangeably.) According to Neil Patel, "Marketers who prioritize blogging are 13 times more likely to achieve a positive ROI (Return on Investment) on their efforts." Blogging also helps to differentiate you and your brand from others in your field. And being more visible can help with sales, too.

Think of your website and blog as the torso (the trunk of the body) of all your efforts, while everything else you create from that blog post becomes the arms and legs (the limbs of the body). You own the torso, but the social media sites you post to, like Facebook, Instagram, and Twitter, own the arms and legs. Now, those sites might say that you own the legs...er, the content...you post, but if you ever get kicked off a site (which happens all the time!), you may not be able to retrieve "your" content. All your hard work could be gone...unless all or most of the content you shared on the social media site originated on your website. That's really the only way to guarantee that your content belongs to you and that you can't be locked out of it.

Describing how to create a website is beyond the scope of this book, but there are many (maybe too many) people who can help you with creating a simple site. Or you can get on YouTube and watch videos to learn how to create your own. You might also want to create a free website, such as Wix, just so you can get started. One caveat, though: free websites usually have lots of ads, so be forewarned. If you think those ads might annoy your potential readers, you might not be wrong, as my brother would say.

Maybe you do have a website but haven't started blogging. Is it because you find the idea of blogging a bit overwhelming? (It's OK to admit it!) But writing a blog post is actually pretty simple; you can write 300 words in about an hour and add an image, and you're done! Then you have content that you can share all over the place. Seems like a pretty good investment of time when you consider that, in addition to ROI and sales, blogging helps your Search Engine Optimization (SEO), too.

If you want to get the most out of repurposing your content, it makes sense to spend the most time creating content on your website, blog, or other online platform where you truly own and control the content, creating the posts that you want the world to see. Then use some of the tips and tricks in this book to get even more mileage from your original writing!

If you don't have your own website yet (because you'll really need one if you want to get the most out of the tips in this book), you can publish in other places, such as Medium, which allows contributors to post anonymously or under a pseudonym. I have friends with their own websites who still choose to publish certain content on Medium. They use it for very personal posts or for other content that they don't feel comfortable posting on their websites (or that, in some cases, their industries might not allow them to publish). For example, certified financial planners are not always allowed to pitch or sell products directly on social media. Some people have to use pseudonyms for their social media posts when they discuss anything even tangentially related to a client or their industry.

There's an upside to anonymity: Being anonymous can give you a secret confidence when you write! And, according to Medium's terms of service, the writer owns the content they publish on Medium.

Pro Tip: Test your blog post ideas before writing them. For instance, if you have two or three different ideas and can't decide which to tackle first, you can ask your friends or followers which they like the best to help you decide. You could even create a short poll on Twitter or Facebook and ask your followers there.

How to Reuse Your Content on Twitter

Now that you have written your first blog post, you can tweet about it and link the tweets to your blog. Here are some tips for the best way to approach using Twitter.

Carol Stephen
@Carol_Stephen

"You might wonder why you stopped doing some things that you really enjoyed. Is the newest, shiniest thing always the best, you might wonder? No, it's not!" When Going Backwards Will Make You Look Forward: ow.ly/gJLK50AvgCa #marketing

Although a tweet can now be as long as 280 characters, don't use all 280 for the body of the tweet. Keep in mind that your Twitter handle takes up some of the space. For instance, my handle, @Carol_Stephen, takes up 13 characters. And you should always leave some space, at least 10 characters, in your tweets so that others can retweet them more easily. For a tweet linked to a blog post, try using a quote from the post, then a headline (just a sentence or two), then a link to the post, and finally a hashtag or two. In the example above, I've shortened the link.

Optimally, use 100 characters or fewer. Sprout Social suggests that the ideal length of a tweet is 70-100 characters. Include a quote, title, or question, plus a URL to your website, and a hashtag or two (at the most). Direct messages can be much longer (up to 10,000 characters!). But don't use direct messages as soon as you get on Twitter or you'll risk being unfollowed or—even worse—blocked!

Headlines and images are really important, so make sure that your headline and accompanying image are both compelling. People will retweet an image they like even without any text!

By the way, I highly recommend the 80:20 rule, also known as the "Pareto Principle." That is, 80 percent of the time you retweet other people's tweets that you think your audience would like; and the other 20 percent of the tweets are yours. I recommend using the "Retweet with comment" feature so that you can include a comment with whatever you're retweeting.

If you've never written a tweet before, get on Twitter and look at some tweets for inspiration. Search within Twitter for tweets that are similar to what you might want to post. For example, if you are a real estate agent, search on real estate, or #RealEstate, and see what comes up. Take a look at the top posts. Do any of the tweets inspire you? What do you like about them? What do you not like about them? Now create a tweet (or two or three) about your subject matter, using one of the tweets you like as a template.

Use a scheduler for at least some of your tweets (Hootsuite and Buffer are both popular). This way, you can plan a bunch of posts all at once and decide when you want them to be posted. People on Twitter generally have a short memory, so you can reuse those "evergreen" tweets (content that stays fresh for a long time) a few times.

For the best results, you'll want to tweak each tweet a bit and repost it at different times. For instance, you may want to use different hashtags and post at different times to see what works best for a particular tweet. If I post my original tweet at 9:00 a.m. on Thursday, maybe I'll repost it on Tuesday at 3:00 p.m. You'll also want to measure the engagement of each tweet with Twitter's own analytics to see what works best. In an article from Social Media Today, Dan Gingiss says, "Yesterday's good content is still good content today, and will be good content tomorrow. It will likely even be good content next week, next month, or next year."

Scroll through some popular Twitter accounts to get a sense of why they are popular. For instance, Popular (coincidence?) Science (@PopSci) recycles content and is able to get more likes and retweets that way. Another example is one of my clients, @WindowWorksUSA. Notice how often something is recycled there?

Pro Tip #1: Save your tweet in a document (Word or Google docs both work well) to reuse it at another time. Tweets don't last very long at the top of people's feeds, so you can reuse that same tweet in a week or longer as long as it's evergreen. Don't reuse the same tweet over and over, back to back, though. That is guaranteed to annoy your audience. Experiment with the best time to tweet for the most engagement. And use Twitter's analytics to see which tweets perform best. Repeat the high performers after you've analyzed why those tweets performed the best.

Pro Tip #2: Put your tweets into a spreadsheet and use a bulk scheduler to republish them (I like Hootsuite's bulk scheduler). This way, you can schedule a week or a month at one time. Just don't forget to get in there and engage, too! Nobody likes someone who just broadcasts all the time.

Pro Tip #3: After you post your evergreen content, you can search your own Twitter stream for it. Then you can reschedule those tweets again.

If You Ignore Facebook, You Might Be Sorry (Create Facebook posts from your website content.)

Although Facebook (deservedly) gets a lot of flak over privacy concerns, there are still more people on Facebook than any other social media platform. According to Statistica, the global number of Facebook users is expected to reach 4.41 billion by 2025. You definitely should not ignore Facebook as a place to publish your content.

Facebook generally allows you much more space than Twitter. A post can include a sentence or two, a quote from your article, an image, and a link back to your website. Personal images and videos do very well on Facebook, with your own content usually performing better than something you've reposted from someone else. Sometimes a very simple question is a good way to generate discussion. I like to start with a "Yes or No: Do you like xxx?" type of post. And images or video nearly always outperform plain text. (More about images and video later.)

Facebook's native scheduler is a great way to schedule Facebook posts. It's easy to do, and you can always change the schedule of posts and/or edit the content later if you like. To make your life easier, schedule a few posts all at once. It's so much easier to use a block of time to schedule your posts in advance. I like to schedule at least three to four posts at once, but having a week of content already scheduled is even better.

You can also use a third-party scheduler, such as HootSuite or Buffer, to schedule your Facebook posts. Back in the day, you would be penalized for scheduling on Facebook using a third-party scheduler, but no more.

Don't forget to reply to any comments that people have made on your posts. That makes your posts perform better and appear to more people. Engagement is what puts the social in social media!

If a particular type of post appeals to your audience, try to find similar content. For instance, if an infographic gets a lot of traffic, look for other infographics you might use. And, of course, reuse that content (just not right away—wait a few days or weeks). Think about why one of your posts might have performed well so the next time you repurpose your content you can recreate some of the magic from the first time.

Note: Hashtags don't generally work very well on Facebook. People have actually done studies on where hashtags work well and where they don't. On Facebook, they don't. If you really love hashtags, then you can go all out with them on Instagram. But you don't really need a study—ask your friends when they last used a hashtag on Facebook. Or ask yourself! When's the last time you searched on a hashtag on Facebook? Most people simply don't use them on the platform.

Pro Tip #1: Include a call to action. This could be to follow you on Twitter, go to your website, or leave a comment. Or, simply, "What do you think?" works well, too.

Pro Tip #2: Use "In case you missed it" (or ICYMI) and then repost the article again on Facebook. You could use a different image with the same article to get even more engagement. And make sure to change the quote, if you used one. You may also want to post at a different time to hit a slightly different audience.

Make Your Post into a Unique and Powerful Podcast (Podcasting can give your article another dimension.)

Podcasts literally give your written content a voice. And we've all heard the saying that people do business with those they know, like, and trust, right? Your voice is another way for people to get to know you. Don't be embarrassed to ask questions about podcasting in case it's new to you. There are plenty of tutorials and articles about how to podcast. For an excellent tutorial on how to create a podcast, see NPR's Starting Your Podcast: A Guide for Students. You may also want to go to YouTube to watch the million or so tutorials on podcasting.

You can add context to your podcast by talking about where you are or including another personal element. Some people have wonderful voices, but, if you're not one of them, you might want to hire someone to create a podcast for you. Keep in mind, though, that quirky and unusual voices appeal to a lot of people. You don't have to sound like a DJ to do your own podcast.

You may discover another point or two while you're recording a podcast that you can use to create a whole new episode. You just never know where the ideas are going to come from, or when. Mine seem to hit when I'm in the shower, but that's a whole other story!

When you record yourself speaking, you can upload that file to YouTube or Podbean. Some people are more auditory, so a podcast can help you expand your audience. You might want to ask someone who's a regular podcaster what kind of special equipment they have, such as type of microphone, to make the sound quality better.

If you don't like the way you look when you're speaking but want to post your podcast to a visual medium, too, you can use a static image with a picture of yourself and the title of your podcast. You can use this same trick on YouTube.

Some people, like Eric Lofholm, my business coach, host live broadcasts where people can call in. They record these broadcasts and post them on YouTube or Podbean. This is a clever way to get more mileage from your posts.

Once you've recorded your podcast, you can post it all over social media: Facebook, Twitter, Pinterest, Instagram, LinkedIn, and anywhere else you have a presence on the Internet. Wherever you post, bear in mind that each social media platform has different kinds of users, meaning different audiences for your podcast. And the time of day you'll get the most engagement will vary from platform to platform, too. For instance, the best time to post on Pinterest, as I write these words, is 6:30 p.m.

Still not convinced that podcasts are valuable? Look at it this way. Not everybody likes to read. Or, if they do read, they may also like to listen. We're so overwhelmed with visual content these days that sometimes it's nice to close our eyes and just listen, especially if you've been staring at a computer screen all day long.

Pro Tip #1: Drive people to subscribe to your podcast, whether you're on YouTube, Podbean, or wherever. That way, they'll receive your newest podcasts when they go up.

Pro Tip #2: Announce your podcast occasionally on all your platforms, and tell people what the newest one will be about. You may want to create a graphic with the details, such as the day or time of day your podcast is usually up. If you have people joining you to record the podcast, make sure to follow up with them about the subject matter and when you're planning to record.

Conquer Pinterest with an Amazing Pin

Pinterest is terrific if your target demographic is women (millennial moms in particular), since 80 percent of Pinterest users are women. According to Sprout Social, 29 percent of U.S. adults use Pinterest (the U.S. is the biggest nation of users of Pinterest). And, according to Pinterest, more and more men are signing up all the time. Another audience segment for Pinterest is people interested in how to do things themselves ("DIYers"). As with other social media sites, you might be surprised by who's on there and what their interests are. So you might want to give Pinterest a try.

I love Pinterest, was an early adopter, and have visited Pinterest headquarters in San Francisco many times for technical talks. Pinterest is a very underutilized site that is terrific for searching for things you might want to make or buy. Also, content on Pinterest lasts 110 days, which is astonishing compared to other sites. Facebook content lasts for 5 hours, and Twitter lasts only 17 minutes, according to ZDnet. Although I'd argue that your pin on Pinterest essentially lasts forever—or until the end of the Internet (whichever comes first).

There are people on Pinterest who might not be on Facebook or on other social media sites, so another benefit of repurposing on Pinterest is that a different audience will see your content. Retailers or Etsy sellers, in particular, find that Pinterest is a goldmine.

Search for others in your own niche to see what kinds of posts they're creating. You can't see their performance, but you can see things like follower count. You may want to follow others in your niche to see what they're posting, too.

The best format for Pinterest posts (called pins) is a longer and narrower graphic than you'd see on other social media platforms. Add the URL of your website, a headline, and hashtags. For hashtags, you can add 8 to 10, as recommended by Pinterest—but no more than 20. Put any hashtags after the description, which has a 500-character limit. Keep in mind that the first words are those that people will see the most, so make those count. And don't forget that most Pinterest users are on mobile devices.

Also, even if you don't have the long, narrow format favored by other people who pin, don't let that stop you from pinning on Pinterest! Just get started.

On Pinterest, you can repin a pin that didn't perform well at a different time or on a different day. You can also repin a pin that did perform well to get more eyes on it. Just as with other social media platforms, don't ever be afraid to reuse your own content!

Pro Tip #1: Add the hashtags after the description. That way, your readers don't just see a bunch of hashtags in the description, which can be distracting.

Pro Tip #2: The best posting time for Pinterest is in the evening, around 6:30 to 7:00 p.m. I've found that the end of the week works well. Of course, experiment to find your own best times to post.

MAKE YOUR POST DRAMATIC TO GET FOUND ON INSTAGRAM

Like Pinterest, Instagram has a younger demographic, although some more seasoned users are joining the fun. After Facebook, Instagram has the most engaged users on social media, with over one billion monthly active users. Again, you'll find people on Instagram who aren't in other places. Instagram, like Pinterest, is very image-centric. People love beautiful images!

Make sure you label your Instagram posts. The label helps people find you, and that's what you want: discovery. You can also use an Instagram story, since that puts your content right in front of people. Even if people don't click on your story, your having a story lets people see you and your brand.

For Instagram, your URL needs to go into your profile since you're not allowed to put the URL in an individual post. So most content creators and brands repeat the message that the "link is in the bio." You can do something similar. Like with Pinterest, I've found that Instagram posts do better when they're posted later in the day.

Instagram is hashtag-crazy. You can use as many as 30 hashtags on an Instagram post! And if you're a brick-and-mortar store with a physical location, use location hashtags. For instance, you could use #SanFrancisco #California if you're located in San Francisco. If you're on vacation in San Francisco and posting vacation pictures, you could also use the San Francisco hashtag. Hashtags are a huge topic, and I can't cover all of that topic right here. But you should definitely include hashtags about your topic and your location, and you can even use snarky ones for fun. For instance, for the Instagram post above, I used the following hashtags:

#SantaCruzMountains, #SantaCruz, #thewildblueyonder, #pacificocean, #viewsformiles #empiregrade, #drivinginmycar, #letsgo, #cloudyskies.

Use your own older Instagram images to create blog entries or Facebook posts. That way, there should be no issue with ownership, since you own those images. And you don't have to go searching all over the Internet for an image for your next post. You might also want to use the image and then tell the story of how you got the shot. For instance, did you use a special tripod or time-delay? Did you set the alarm to get up early to capture the shot? Do tell!

Pro Tip: Create a custom, branded link for Instagram, and put that in your profile for each post. You can use the custom link on Twitter, and on LinkedIn, too. I like bit.ly for branded links, which are also a good way to track engagement. For instance, here's a custom link that goes to my author's page on Amazon: http://bit.ly.CSAuthor. You could also put the subject matter and/or the date in the link so people know what they're clicking on.

SHARE YOUR PROFESSIONAL WRITING ON LINKEDIN

LinkedIn has the most professional, and most business-minded, people on it, with many using it to find good content and also to increase their digital footprint. LinkedIn is perfect for business-to-business brands and for finding collaborators for your business ideas. If you've created a post, you can share it on LinkedIn, but make sure to share it at a different time than on other platforms. This is a good idea in general. People don't have an incentive to follow you if they see you blasting identical content on multiple platforms all at the same time.

Recently, LinkedIn merged with Microsoft and has also acquired Lynda, the learning platform. With over 630 million users worldwide according to Omnicore. LinkedIn is a sleeping giant.

As on other social media platforms, timing matters on LinkedIn. Recently, at a business conference, I heard some very good advice about timing posts on LinkedIn from a LinkedIn executive. Sharing short articles is a good idea, especially early in the morning or after dinner. Try posting at 7:00 a.m. and after 6:00 p.m., and see what happens.

You can share your articles more than once on LinkedIn, but let some time go by between the times you share, and preface these posts with "from the archive," "Throwback Thursday," or "In case you missed it."

On LinkedIn, as well as on most other social media platforms, it's good to use the Pareto Principle (the 80:20 rule that I talked about previously). That is, you share 80 percent of your content from others, and share just 20 percent of your own content. If you post five days a week, you could pick one day to share your own content.

Bear in mind that commenting on others' posts is a form of content, too. People see the comments you make, as well as which posts you "like."

LinkedIn isn't just for recruiters. If you have solutions for businesses, then your content belongs there. You can create short-form content on LinkedIn, along with a link back to your website. I also like to let people know if content has been previously published (if you're sharing such an article). Original content will always do better than resharing your previously published posts, but LinkedIn is still a great place to repurpose your content.

As a rule, hashtags don't do well on LinkedIn, although LinkedIn has recently been suggesting them. I'd avoid them, if possible, for the time being. Also, I'd use full sentences, rather than short phrases. Assume that your audience will read your post all the way through, especially if the post is something that might help them with their business.

ADD A FUN AND DIFFERENT IMAGE WHEN YOU REPURPOSE CONTENT

Approximately 65 percent of people are visual learners, so you should always include an image in your post. For many people, the image is the most important part of your post. Wouldn't you rather click on an article if it has an interesting image? If you're like many, when you write a blog post, you include more than one image. So choose one of the other images when you repurpose that content. Chances are, most people won't remember your original post, or they'll get something different from it when you share it again. When you reshare a post, you may also want to use a different quote if you used one the first time you shared.

Wait a few days before you share the post again. Or you could wait a few weeks or months, too. This is where creating evergreen content really pays off. If you're sharing a post about an event, you can generally only use that content once. If you repeat the event, you can look back at previous years' images, though, and possibly alter those. And you can use the #ThrowbackThursday or #TBT hashtags if you want to reminisce.

If you really like images and have a few associated with a blog post, you can add more than one image when creating a post on social media. For instance, on Facebook, three images together look better than two images together. Upload the images one at a time, before you upload the text and URL—otherwise, Facebook will usually populate the post with whichever image is at the top of your post.

By the way, you can use a free image site, such as Pixabay, to get free pictures to add to your post. Unsplash is another favorite for free images. But make sure that the image says that it's for commercial use. Otherwise, you could be sued or charged thousands of dollars for stealing.

Pro Tip #1: Take a screenshot of the page where you got the free image, and make sure the screenshot shows that the image is available to use for commercial purposes. Trust me, this can save you a lot of headaches down the road. Rename the screenshot so you can find it easily later, and save the image in a folder you use just for these sorts of images. A photographer or graphic designer may change their mind about that photo you used being free, and you could get in trouble if you're not careful to follow the above steps.

Pro Tip #2: Repurpose what you share on social media to an email newsletter. Email newsletters may have a different audience than your social media does. Many people who don't use any social media sites at all will read an email newsletter.

CREATE A COLORFUL VIDEO FROM YOUR ORIGINAL CONTENT

A video, like a podcast, can be used again and again on its own. Some people prefer watching videos to reading. And, again, people want to feel comfortable with you before they buy from you. So, letting people hear your voice and see your face are great ways to introduce yourself to your audience and make them feel more comfortable with you. A video can be a more personal way to share. As with a podcast, you can share where you are as you're making the video, what your day has been like, the weather where you are, and so on. And if you're shy, you can make a video using your voice, but be behind the camera, so to speak.

When you create a video, you'll probably find yourself speaking a little differently than when you write. When you speak, you're probably more casual than when you're writing.

A video of one to two minutes is the best for social media posts. Of course, if you engage in storytelling, a video can be longer, since people want to hear the end of the story. Time seems to stand still when you're telling a story, as opposed to simply doing a "how-to" video. You could even start a story in one video and then continue it in the next video. For instance, I'm a sucker for stories about animal rescues. Someone finds a bunch of kittens hiding in a sofa, rescues them, and you watch them heal. Don't you want to know the end of that story? I do! As you're telling your story on video, you may come up with even more ideas for other videos.

Keep in mind that Google owns YouTube, so your success on YouTube helps you on Google, too. The other thing about YouTube is that you never know when one of your videos could be "discovered." I have a video with over 100,000 views about organizing jewelry from back when I was a professional organizer. But it didn't happen overnight!

And now that Facebook—as well as other platforms—supports video, you can use the native video feature to go live, which is a quick and easy way to make a video. To figure out the best time to share your video, you may want to look at your analytics on Facebook (or wherever you're making your video) to see when most of your users are on Facebook. And little snippets of video do well on Instagram and on Instagram stories, too. You can do video on LinkedIn, and even Pinterest is getting into the video game.

WHAT HAPPENS WHEN YOU CREATE AN INFOGRAPHIC?

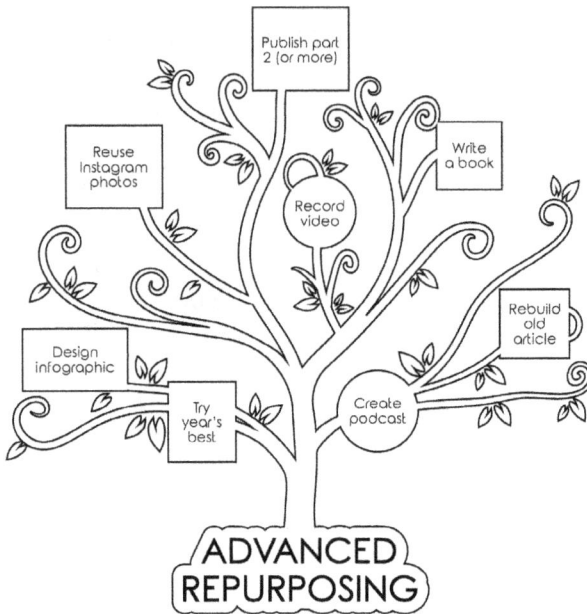

At this point you might be wondering, why use an infographic? (Some folks may actually be wondering, what is an infographic? An infographic is data organized in visual format.) So, why use an infographic? Infographics appeal to different types of people. People with an analytical mindset love infographics. When I took a course on Pinterest, the instructor mentioned that infographics are a way to engage male users there. Infographics give a lot of information at a glance, especially factoids that might be difficult to relay in purely text format. You can include graphs and charts in an infographic, too.

One way to make an infographic is to use Canva, a graphic-design-tool website. There are other sites that can help you create infographics, too, such as Piktochart or Venngage. You can also hire someone on Fiverr or another site to help you create an infographic. Infographics tend to appeal to more scientific minds.

An infographic is a way to make otherwise dry information more interesting. You can use different colors, textures, and maps for your infographic. And since a lot of people don't like to read, one picture really can equal a thousand words. You can emphasize your most important points using color, size, or orientation.

Need some more examples? Go to Pinterest and search for the word *Infographic*. Then use Pinterest's guided search to narrow your search for the topic you're interested in.

Now that you have an infographic done, you can share that infographic all over, from Facebook to Twitter to LinkedIn to Pinterest. And, if you haven't already done so, write an article around the infographic, too. Make sure to put your own brand name on your infographic in case someone decides to "borrow" it and use it as their own.

Pro Tip #1: Spend some time planning how you'll share your infographic. When will you share it? What will you say about it? Will you share the entire infographic, or will you have a link back to your website for the full infographic? And, of course, when will you share it again for maximum impact?

Pro Tip #2: Since Pinterest users like long, skinny graphics, Pinterest is a great place to share your full infographic, unlike other social media sites.

THE BEST WAY TO TAILOR YOUR CONTENT FOR EACH SOCIAL MEDIA SITE

Each social media site has its own language and rules. Of course, you can break the rules, but you'll get more engagement as a general rule if you follow the rules and use the language of each site. Starting with one article, you can make it look appropriate for each platform. This really is pretty easy—I promise!

Here are some tips to tailor your content for a few of the top social media sites:

Facebook: Create a short post with a question format, one or more images, and a URL if you want to send people to your website. Give them a call to action (buy my book, click for more information, leave me a comment, etc.) for each post. Hashtags really don't do well on Facebook, but if you really want to use a hashtag, put it at the bottom of your post. There are some exceptions to the hashtag rule, though. For instance, if you have a series of articles, you can use hashtags to help readers find the articles more easily. Another exception might be if you have a branded hashtag that your followers associate with your brand.

Twitter: Headlines and images are really important, so make sure your headline and accompanying image are both compelling. Use 100 characters or fewer. Use about 100 characters for a tweet. Have a short quote, a title or question, a URL to your website, and a hashtag or two (at the most). By the way, direct messages can be much longer (10,000 characters!).

Pinterest: Use a longer, thinner graphic, a title for the pin, a full description of the pin, and 8-10 hashtags. When people first started using Pinterest, many posts would just have a period (".") in the description. Don't do that! *Think like a search engine and put some descriptive words in the description.* If you were looking for your post or article, what would you type into Google? Type that into your description on Pinterest. Put any hashtags after the description. And remember that Pinterest's character limit is 500 for a description and that most Pinterest users are on mobile.

Instagram: Make sure your graphic is beautiful. The language people use on Instagram varies but is generally more casual than on other platforms. Have a description of the post, plus 10 or more hashtags, including geographic hashtags like your city name. A hashtag is how people find you. There are lots of accounts with no hashtags and they generally have very few followers. If you have a brick-and-mortar store, location-based hashtags are even more important. You can use up to 30 hashtags on an Instagram post.

LinkedIn: Use full sentences with good punctuation and grammar, and don't use hashtags. Like everything else on social media, that rule could change at some point, though. You may want to use a little more text, but don't use slang, if possible, on LinkedIn. The demographic is a more serious group, and users probably won't appreciate it.

For all social media sites, make sure that you're engaging with your followers. If engagement is difficult for you, check out my previous book on engagement, 21 Ways to Social Media Engagement: That Will Make You Look Like a Pro.

A NEWSLETTER CAN MAKE YOU RETHINK YOUR CONTENT

You've already spent a lot of time and effort to write your content for your website, blog, and other platforms. So why not create a newsletter from it? Some people will read a newsletter before they'll read something on Facebook or Twitter. Depending upon the age of your subscribers, they may not even use social media. There will always be people who aren't on Facebook, Twitter, etc. Most everyone has email, although many younger users don't check it very often.

Pia Silva, in her Forbes article *Why Newsletters Suck and How to Do Successful Email Marketing* suggests that as you consider building your own email newsletter, you should think about the emails you get that you do enjoy. For instance, she does not enjoy the email that Netflix sends out, but I do. I like the suggestions on what to watch next as well as the graphics. Think about what you can replicate about the emails that you continue to subscribe to and enjoy.

Make sure to include good images in your newsletter. Images are great to use no matter where you're repurposing your content, but they particularly lend an emotional edge to a newsletter article, which could otherwise come across as dry or formal.

You may want to add an introduction to each piece of newsletter content, such as the context for each post. Include a link back to your website if you want people to read more or you want to drive more traffic to your website. Also, keep in mind that your life or the particulars in an article may have changed between when you first wrote an article and when you crafted it into a newsletter. Make sure to use a shortened link, as they look so much neater in a newsletter. You can use bit.ly for this. And you can even make a custom link, which helps you track where your leads are coming from.

Most newsletters have a number of stories in them, so, if you're a regular writer or blogger, you could use two or more shortened articles with links back to your website. For instance, I usually write four blog posts per month, so I could put two blog posts in each newsletter and send it twice a month. Other newsletters just duplicate the website articles and don't include links back to the website at all. However, this usage doesn't really give you the benefits of repurposing content, which, after all, is the point of this book.

If you're using your newsletter to link back to the articles on your website (and I hope you are!), include an introduction or other intriguing text for each article in the newsletter that tells people why might want to read the article so they are more likely to click the link. Some people also tell their readers how long an article will take to read, too.

If you're not good at writing newsletters and are not interested in learning, there are lots of people out there who'll help you. After doing a quick search on Fiverr, I found all kinds of people who can create an email template for a newsletter or even help with the design.

The two email services that are most popular for newsletter creation are Mailchimp and Constant Contact. Constant Contact has more features, but Mailchimp is more affordable.

How to Reuse Your Best Instagram Photos in Other Places

It's easy to edit a photograph in Instagram or on your phone using filters. And phone cameras are becoming more and more sophisticated. While you're repurposing the text you previously wrote, you can also repurpose the pictures you took. Take one of your Instagram photos and, using a program like Canva or Photoshop, you can write the title of your post across your image.

Now take that image and use it all over the place. Since it's your photo, you don't have to worry about any copyright infringement issues. Did you know that if you don't have the rights to use photos, you could be fined for using them illegally? I've heard of fees of up to $1,500 for using a single image. Do not take images you find on Google searches or on Instagram and use them as your own. I've had clients do this before, and they're likely to get caught and fined if they're found out! And, of course, it's stealing.

Check on YouTube for easy ways to edit your Instagram images. Many times you can edit that photo right on your phone. And you may not even have to use an app, since there are so many filters right on Instagram itself! Personally, I like to warm up my photos and change the saturation a little.

There are lots of different apps for different experience levels. You don't have to be Ansel Adams (or Peter Parker!) to come up with great photos to post.

You could also find someone on Fiverr to create or alter an image for you. But I'd recommend that you check YouTube first to see if there's a short and sweet video that would help you instead. Some of the best images make people want to see what's behind them. There are quite a few videos on editing your Instagram photos in Canva, for instance.

Pro Tip #1: On some platforms, it's better to have just a small amount of text on an image, so you may want to create multiple images. For instance, if you're going to do a Facebook boost or ad, an image with a lot of text won't be seen as much as an image with less. And some Facebook images can be rejected as ads for having too much text in them.

Pro Tip #2: If you're a good photographer, why not use your Instagram photos to create a calendar for your clients?

WHEN YOU REBUILD AN OLDER POST, YOU CAN HIT THE JACKPOT

Sometimes, despite your best efforts, a post doesn't get any traffic. You spend a lot of time creating it, you choose the best images to go with it, you post it at just the right time, and...crickets. So why not take that underperforming post and see if you can turn it into a popular post? Maybe all it needs is a different title and a jazzier image (or two). Just like with a house remodel, you probably won't change some or even most of the house (the post), but you'll most likely want to update the appliances (the images) and add a new roof (that would be the jazzier title).

It's good strategy to try to figure out why the original post failed to get traffic, too. Was there a competing event that distracted people at the time you posted your article? Was the call to action too complicated? Did you post at an odd time of day or on a less-than-optimal day of the week? Maybe the image was inappropriate or not emotional enough. You could also ask a friend or two what they thought of the post you created and what they think could improve it.

Analytics can only tell you so much about your posts and why they succeed or fail. A large part of your success on social media is gut instinct and guesswork (no matter what "experts," even those who have written books on the subject, may tell you). I hardly ever know which posts will really do well, for instance. I can guess, based on my own emotional response, but sometimes a post will take off and

I won't know why. Other times, I see a video and know that my friends and followers will love it.

You're not really creating something from nothing, but you're tricking yourself, in a way, into writing something that's new, starting with a small seed of an idea. Do you remember the folktale of stone soup? It's a lot like that. Once you have that first stone (or seed of an idea), you can add another idea, then another and another, until you have a whole new post. And, of course, you can then repurpose that post, too.

Pro Tip #1: Create an entirely new headline to go with your post. The headline needs to be enticing enough so that it will compel people to want to interact with it. Consider what headlines you respond to. Could you write something similar? By the way, CoSchedule has a fantastic tool for coming up with great headlines.

Pro Tip #2: Take an older post that featured items in a list and expand on each of the items. For instance, "Ten Ways to Better Real Estate Transactions" could become 10 different articles, each with a link back to the original post.

Share a Bit of Yourself as Context for a Zinger of a New Post

Sometimes you'll find yourself writing a post on a topic that has, frankly, been done a million times before. If you can add some personal details, especially about what prompted you to write the post, you give your readers a more compelling reason to engage with the post.

For instance, maybe you're writing a post about relaxation. Think about including your own struggles with insomnia or other sleeping difficulties as context for the post. When you share the post, also share some of the details about how you came to want to write on that topic.

Your readers will appreciate knowing that you, too, have horrible clown-themed nightmares just like they do! They may relate personally to being chased into an alley filled with disappearing cars that honk incessantly!

Or you might want to talk about the ways you have found to relax, perhaps to inspire them to do the same. People love to have the context of a story so that they can feel more personally invested in and connected to it (and, by the transitive property, to you). Sometimes they'll see themselves reflected in the more personal details, and that will make them want to engage with your post more.

Often, I see posts with no context at all on social media. I tend to skip these posts. Let us into your world. Tell your readers why you like what you're posting, or why you wrote something, or anything rather than just a picture.

And, if you really believe the image speaks for itself, you could just say "No words." Sometimes that works.

People don't always remember to consider their audience. They write and post the first thing that pops into their head without including any context. These types of posts can seem like inside jokes, and they tend to alienate rather than engage followers or friends. For instance, during football games or the NBA playoffs, people will just say something like "Way to go, Steph Curry!" Don't do this. People who don't know anything about basketball will be left wondering what you're talking about. And some of them may mute or even unfollow you! That's no good.

Pro Tip: You may want to comment on your own post. Sometimes people do this with great success. While this isn't repurposing, exactly, it does give your post a little boost. I've seen people do this with success on Instagram. Definitely answer comments with your own comment to help a post gain even more momentum.

Curate the Year's Best Posts with a Tantalizing Summary Post

If you post a lot on social media, you might want to do a best of the week post or a best of the month post. You could also collect some of your best tweets, along with your friends' tweets, from a Twitter chat. This is a good way to include a lot of people in one post, as well as a way to create content.

And people love to be included in your articles. Don't you love to be included and given credit for something you said or did? I know I do! People who are included tend to comment more on posts, too. Using quotes from friends and followers lets you create more content, which you can then reuse again and again. You could say, "Linda and Harriet use this method all the time," while "Lisa and John never do," or words to that effect.

At the end of the year, you could summarize what you've done that year on social media and maybe talk about what has and hasn't worked to give your followers a glimpse behind the curtain. This is a good way to recycle some of your posts while giving your followers something new at the same time. It's also an opportunity to review the year's posts and maybe learn from what you've done, too. Some good titles could be "What I've Learned from 12 Months of Blogging," "My Favorite Posts of the Year," etc.

Pro Tip #1: Ask your readers what they thought your best post of the year was. You might also ask what they'd like to read about for the new year or if there are areas where they feel you could improve. Most people love to give advice. Although this might leave you a little vulnerable, it's a good way to gauge what your audience finds important.

Pro Tip #2: On some platforms you can create a poll. (I often use Twitter polls to see what friends/followers are thinking.) Then you can include the results of that poll or post it in other places. For instance, you can take a screenshot of the post from Twitter and post it on Facebook. I do this sometimes during a Twitter Chat that I run, and then I post the results on Facebook.

SEE WHAT "THROWBACK THURSDAYS" CAN DO FOR YOUR OLD POSTS

Throwback Thursday is an institution on social media where people post things they want to revisit. This is the perfect chance for you to brag a little (or a lot) about something that you wrote.

For instance, I recently posted a picture of myself at Facebook headquarters from years ago, with the caption: "Remember when Facebook's logo was red?" (Most people didn't, by the way.)

And, with Covid-19, many of us measure time differently. There's the before time and the after time. Hopefully, by the time you read these words, we'll be well on our way to a vaccine.

If a post didn't get much attention the first time you published it, why not use Throwback Thursday as a way to get more eyes on it? There are some generally accepted rules for Throwback Thursday, so keep them in mind before you go posting something from last week. A post or picture should be roughly five years old. The picture could be from before the platform even existed. But maybe also something you wrote that wasn't in blog format, for a class or something, could also work. It doesn't necessarily have to be five years old if people haven't seen it before.

Throwback Thursday has its own hashtag: #TBT. And, according to the Throwback Thursday Wikipedia page, the phrase has been used over 200 million times! And its usage is still spreading. Generally, nostalgic photos or videos do best on Throwback Thursday posts.

By the way, if you like Throwback Thursday, there's also Flashback Friday. Apparently, Flashback Friday was popular before Throwback Thursday. However, Flashback Friday didn't take off the way Throwback Thursday did. It's one of those chicken and/or egg things, I guess.

Facebook has done an impressive job of getting us all to look back and reshare some of our own content. It certainly makes their job a lot easier, too! Think about how we are the creators on Facebook. The engineers at Facebook want us to stay on the platform longer, and having us share more is a good way to do that! Very clever, Facebook!

Pro Tip: If you're reusing content and someone else is in that picture or video you share, make sure to tag them so they see it, too. Don't rely on people checking your page or account and seeing what you do—tell them!

SECOND VERSE? NOT THE SAME AS THE FIRST!

Create a Part 2 for a follow-up from your most popular post or posts. If your post is too long, split it in half. You may find yourself with more to say about a previous post. You may want to create a Part 3, or even a longer series, if you find that you have a lot to say about a particular topic. Think about some of your favorite blockbuster movies and how there is often a sequel (or two or three or six).

Also, since people's attention spans seem to be getting shorter and shorter, you want your posts not to be overly long, in general. So, a Part 2 or Part 3 is perfectly acceptable. If you have a 700-word article, cut it in half and add a new photo to the second half so you can get two posts out of it.

In my case, since I write about social media, things change really fast. For instance, as I'm writing this, Facebook's Mark Zuckerberg has been in the hot seat because of Facebook mishandling political posts. Many people are deleting their Facebook accounts because data privacy or politics, while others are canceling ads to the tune of billions of dollars! Social media consultants are waiting for the fallout, with some leaving the Facebook platform to focus their energies elsewhere. So I could revisit a previous post about Facebook with some of this information.

Maybe your business doesn't change as quickly, so you wouldn't need to be as concerned about such posts being evergreen. But take a quick read of your own post before you repost it, just in case some of the information is out of date.

As I mentioned earlier, you want to make your posts as evergreen as possible. For me, I might mention the political situation, but not dwell upon it, as within a few months, people will likely be interested in something else.

Pro Tip: If you're going to repurpose a post that's not evergreen, you could always say "Remember when everyone lost their passwords during the Cambridge Analytica breach?" And then do a follow-up article on it. Or something that applies to your business.

THREE OUT OF FOUR SOCIAL MEDIA MANAGERS AGREE... (TURN A SURVEY INTO AN ARTICLE.)

You could email a survey to some of your clients, or create one on social media, and use that as a jumping-off point. Record your reaction to the results of the survey. Were you surprised? Or did the results serve to underscore what you already knew?

When you post your article, you might want to tag some of the people who responded to your survey, too, for additional reach. Use Google to check some of the best ways to create a survey. I like Survey Monkey for quick and easy surveys.

For instance, I recently asked people in a Twitter Chat what subject they'd like us to cover the following week, and a lot of them said they were interested in learning more about analytics. So, in our next #DigiBlogChat, we tweeted about analytics. Before I finalized the topic, I went into our private Facebook group to ask what, specifically, people wanted to learn about analytics, since it's a pretty big topic. (By the way, asking the crowd for input is a very good way to find a popular subject.)

The analytics chat was a great success! And, since analytics was such a popular topic in general, I'm keeping all the questions for reuse on another platform and might reword them slightly for a future Twitter Chat (perhaps in a year or so). You can absolutely do this when creating content to reuse.

Carol Stephen
@Carol_Stephen

Q1. How often should you share your own content on LinkedIn? #DigiBlogChat

38% As often as possible

54% Occasionally

0% Never

8% Use an 80/20 strategy

13 votes · 4 days left

Facebook and Twitter surveys are easy to create, and you can use a screenshot of them as the basis for a new piece of content. Note: On Twitter, you can only have four choices in a survey. But you could do a Part 2 of your survey and add four more choices that way, if needed. As an example, above is a survey about sharing content on LinkedIn that I created. With this survey I created on Twitter, I could write a blog post on LinkedIn and share it all over social media.

Pro Tip: If you're using your own feelings or opinions about something as the basis for your writing, chances are the content will be original, something which Google rewards you for. So, a deeper dive into how something affects you can be surprisingly relevant. And you may even want to revisit your own feelings about a topic and how those feelings have changed over time.

REWRITE BEST-PERFORMING POSTS AND YOU'LL LIKE YOUR RESULTS
(GO BACK TO THE DRAWING BOARD.)

First, figure out which post or posts perform the best. There might be one or two or a whole bunch. For this, take advantage of each site's analytics. If you're looking at your website, use Google Analytics to see what's performing the best. You could also use the number of comments you get on a post, if that's how you like to measure your success. Or you can use the analytics available to you on each platform. For instance, Twitter's analytics can tell you which of your posts is performing well.

Create a new headline for each of those posts, and make sure each headline gets a higher score than the headline on the original post. You can use CoSchedule's headline analyzer for this. While you're at it, use a different image for the new version of the post. Ask yourself a few questions about the post: Are the outward links still relevant? Can you reuse the call to action (or make it even better)? What part did the timing play (when it was first posted) in its success? Sometimes a post will catch fire more easily when it's posted at the perfect time. And the perfect time may vary depending upon which platform you're using. The big question, of course, is are you solving your readers' problems?

As you're writing or rewriting your posts, think about where else you'd like to share that same content. Are there some new places you'd like to share? Maybe you're now on some platforms that you weren't on when you wrote that post originally. Maybe you didn't share it on Instagram, for instance.

You can also use this strategy of rewriting best-performing posts on your blog. Which article got a lot of traction online? Could you rewrite it, use different images, and then publish it again as a different article? If there's a lot of interest, perhaps it's a subject that people really crave.

Pro Tip: What if that post that went viral could be followed by another viral post? Really study it to see if you can do the same thing again. You got all the stars to align once, so it's certainly possible!

This Is What Happens When You Write a Book

Here's a little secret. I started writing this book (that you're reading now) before my first book was done, as a way to trick myself into finishing the first book! I began my first book by taking a bunch of my blog posts and rewriting them, editing them, and compiling them into a book. This book is all original content. That said, you might see some of the same themes in the book you're reading right now as there are in my blog writing.

I'm not writing about creating a book to brag, but to let you know that you could do the same thing. Maybe you don't have a bunch of blog posts, but you've certainly had lots of conversations or perhaps written something for a class that you could repurpose. And repurposing is the name of the game.

A book doesn't have to be a huge treatise of 10 million words. It can be a marketing piece, a long pamphlet, or an instructional guide of just a few thousand words. You may feel a bit intimidated by the idea of writing a book, but believe me: if I can do it, you can do it, too.

Once you have your book done, of course you can post, tweet, or otherwise share pieces of the book in other places on the Internet. And share again and again. Why, I know someone who wrote a book in 2014 and her Twitter feed is still all about that book.

Even though I'm (probably) not going to get rich from this or my other book, it's nice to get a little money from Amazon every month. In addition, a book is a great marketing tool. You can give it away to attract clients or email subscribers, or simply use it to advertise yourself as an expert in your field.

So, as you can tell, repurposing is an idea that I really take seriously myself. And will I take pieces from this book and create posts for all the various social media platforms? You betcha!

Pro Tip #1: If writing a book is something you're serious about, why not get into a writing group so you can get support from and share the experience with other people? Whichever way motivates you the best is the way to go. If you like in-person groups, then check online for meetups in your area, or start your own writing group. If you're self-motivated, set a goal to write a certain number of words each day.

Pro Tip #2: Publishing changes all the time. Check YouTube for videos of how to finalize your book. You can find practically anything on YouTube!

NEXT STEPS: CARVE OUT TIME TO IMPLEMENT SOME OF THESE IDEAS

I hope that you have enjoyed this book and that you'll be inspired to recycle more of your content online. If that is the case, then I'll know that I have met my goal. Make sure to put some time on your calendar so that you can put these ideas into action.

Break your bigger goal into smaller goals, and put one on your calendar every day. Even if it's only 15 minutes a day, eventually you will have achieved your bigger goal.

Please connect with me, and, if you need help with your social media, shoot me an email at carol@yoursocialmediaworks.com. Find all my social accounts at Your Social Media Works.

www.ingramcontent.com/pod-product-compliance
Lightning Source LLC
Chambersburg PA
CBHW050522210326
41520CB00012B/2407